Paulo Sergio de Jesus Vilela

Methodology for controlling and automatically directing solar panels

Paulo Sergio de Jesus Vilela

Methodology for controlling and automatically directing solar panels

Renewable energy, the future of the energy matrix

ScienciaScripts

This book is a translation from the original published under ISBN 978-613-9-64134-5.

Publisher:
Sciencia Scripts
is a trademark of
Dodo Books Indian Ocean Ltd. and OmniScriptum S.R.L publishing group

120 High Road, East Finchley, London, N2 9ED, United Kingdom
Str. Armeneasca 28/1, office 1, Chisinau MD-2012, Republic of Moldova, Europe
Printed at: see last page
ISBN: 978-620-7-72797-1

SUMMARY

Due to the increase in demand for electricity and the imminent risk of fossil fuel shortages, as well as the threat of a constant water crisis and pressure for measures to control consumption, it is necessary to develop and apply methods for generating clean and renewable energy, such as solar energy. Even though photovoltaic solar energy is not very widespread and does not have massive investment in our country, there is considerable growth compared to other renewable energy sources. Grid-connected photovoltaic systems are increasingly used throughout the country and regulatory resolutions have been created to support the use of these systems. The aim of this work is to carry out a preliminary study into the possible future application of a photovoltaic energy generation system in the new Interdisciplinary Bachelor of Science and Technology (BICT) building at the Federal University of Maranhão (UFMA). A literature review is carried out in order to design and develop a prototype for the photovoltaic panel positioning system using the Arduino microcontroller in order to increase the system's efficiency.

Keywords: **solar energy, photovoltaic, positioning system, microcontroller, Arduino.**

Summary

CHAPTER 1

INTRODUCTION

Professor Adriano Pires, from the Federal University of Rio de Janeiro (UFRJ) and director of the Brazilian Centre for Infrastructure, is categorical in stating that in recent years Brazil has experienced the biggest energy crisis in its history, which has brought two serious problems: the financial one, created when the government renewed concessions, forcing companies to reduce tariffs at a time when the cost was rising worldwide, and the exaggerated incentive to consumption, causing supply problems (O GLOBO, 2015).

The country's energy situation is a cause for concern, with reservoirs operating at low levels, power stations at a standstill for construction and the possibility of new costs for consumers. The discussions, however, have adopted more of a political bias than, in fact, a commitment or concern with diagnoses and solutions to stabilise the country's energy supply. Among the solutions they present is one aimed at rational consumption, according to Gilberto De Martino Jannuzzi, associate professor at the School of Mechanical Engineering at the University of Campinas (Unicamp). He emphasises that the growing demand for electricity in Brazil could take on unsustainable proportions and that no new hydroelectric power plant would be able to supply it (JORNAL DO BRASIL, 2015).

A good alternative for minimising these problems is the **so-called "alternative energy sources"**, which are gaining prominence in debates on the subject. As well as drastically minimising the damage caused to nature, these alternative sources are renewable and therefore perennial. Examples of renewable sources include solar energy (solar panels, photovoltaic cells), wind energy (wind turbines, wind turbines), water energy (water wheels, hydrokinetic turbines) and biomass (plant matter).

Brazil is a country rich in natural resources and has human resources available to work in photovoltaic solar energy generation. However, despite notable efforts in some renewable energy sources, there are few results that promote the insertion of photovoltaic energy in the National Electricity Matrix.

The Brazilian state is in the process of defining public policies to encourage or regulate the inclusion of this energy source in rural and urban electrification concession networks, as shown in Figure 1. This hampers the development of photovoltaic energy on a larger scale and emphasises the importance of applying regulatory mechanisms to promote business and encourage technological innovation.

Figure 1 - Photovoltaic panels for generating electricity installed on the roof of a house.

Source: Shutterstock (2014).

1.1. GENERAL OBJECTIVES

The aim of this work is to study photovoltaic solar energy, its concepts, applications and feasibility, based on a literature review, for the future sizing of a photovoltaic energy generation system for the new Interdisciplinary Bachelor's Degree in Science and Technology (BICT) building at the Federal University of Maranhão (UFMA).

1.2. SPECIFIC OBJECTIVES

In view of the above, the specific objectives of this work are:

1- Verify the potential, concepts, obstacles and benefits of photovoltaic solar energy.

2- Know the basic concepts and variables that influence the system.

3- Present the application of a photovoltaic mini-generation system.

CHAPTER 2

LITERATURE REVIEW

2.1.　　　SOLAR ENERGY AND ELECTRICITY

The growth in electricity consumption around the world is notorious. In 1980, the world consumed around 7,000 TWh (terawatt hours) or 7,000,000 GWh (gigawatt hours) of electricity and according to forecasts by the International Energy Agency (IEA), this figure is expected to rise to 30,000 TWh by 2030. 1 TWh is equivalent to 1,000 GWh or 1 million MWh (megawatt hours). The average monthly consumption of Brazilian households is approximately 500 kWh (VILLALVA, GAZOLI, 2012).

The photoelectric effect was first observed by Heinrich Hertz in 1886. Hertz accidentally discovered that when a beam of light was shone on a metal plate that was close to another of different potentials, an electrical discharge occurred between the two. By reducing the incidence of light on the plate, Hertz observed that the discharge decreased significantly (FOWLER, 2007). The theory behind the photovoltaic effect was that the energy coming from the incident light literally pulled a number of electrons out of their orbits and expelled them from the material. The nearby plate, with its lower potential, attracted these electrons, thus forming a discharge.

Due to experiments carried out in this area, in 1905 Albert Einstein suggested that in some circumstances, light behaved not like a wave but like a particle, giving it a dual wave-particle nature. This work, which won him the Nobel Prize, is based on the idea that light can be thought of as a stream of particles, called photons, each of which functions as a small packet of energy. The energy corresponding to each of these photons is directly proportional to the frequency of the photon wave (RESNICK, 2002).

Any body at a temperature above 0 Kelvin (approximately minus 273 °C) loses part of its internal energy in the form of thermal radiation. The mechanism behind this emission is associated with the energy released as a result of oscillations or transitions of the object's constituent electrons. These oscillations are in turn sustained by the internal energy and therefore the temperature of the material (INCROPERA and DEWITT, 1996).

The sun can be used to produce electricity, according to Borges Neto and Carvalho (2014), which is the major primary source of energy on planet Earth. Every year the sun provides the Earth's atmosphere with around **1.5125×10^{18}** kWh of energy. This radiation is important as a source of energy for various atmospheric and terrestrial phenomena, such as the formation of high-pressure zones, responsible for wind flows, ocean currents, the evaporation process and others. The sun's energy is indirectly responsible for most of the energy available on earth (ANEEL, 2007).

Photons are found in solar radiation and comprise all the light intensities related to spectral distribution. Outside the atmosphere, the solar spectrum is similar to that of a black body at a temperature of 5,800 k and the irradiation value is 1,367.5 W/m² At sea level, this spectrum is modified due to atmospheric absorption and the irradiation decreases to a value of approximately 1,000 W/m2 (LORENZO, 1994).

Measuring solar radiation, both the direct component and the diffuse component on the earth's surface, is very important for studies related to the influences of climatic and atmospheric conditions. With a history of these measurements, the installation of thermal and photovoltaic systems in a given region can be made feasible, ensuring maximum utilisation throughout the year where variations in the intensity of solar radiation undergo significant changes (BORGES NETO and CARVALHO, 2014).

According to Cresesb (2006), the first reports of solar energy being transformed into electricity date back to 1939, and the first photovoltaic device was assembled in

1876. However, it wasn't until 1956 that industrial-scale production began.

Solar energy can be used to produce electricity through the photovoltaic effect, which consists of the direct conversion of sunlight into electrical energy. Unlike solar thermal systems, which are used to heat or produce electricity from the sun's thermal energy, photovoltaic systems are capable of directly capturing sunlight and producing electric current. This current is collected and processed by controllers and converters, and can be stored in batteries or used directly in a system connected to the electricity grid. (VILLALVA, GAZOLI, 2012)

According to ANEEL (2012), grid-connected photovoltaic power generation systems have been the fastest growing electricity generation technology in the world in recent years. This is due to a change in the behaviour of enterprises, which a short time ago were mostly intended only to serve isolated regions without coverage by the conventional electricity distribution network. And as the application of this technology becomes more widespread, its cost tends to fall, as shown in Graph 1.

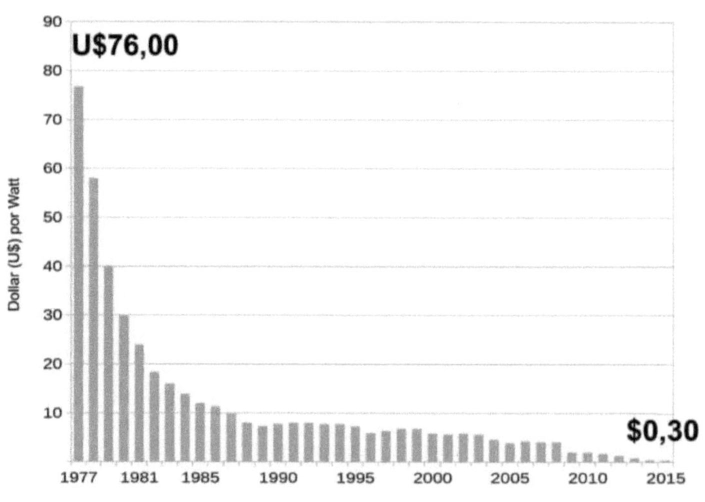

Graph 1 - Values of photovoltaic cells watt/dollar.

Source: Solar Portal (2016).

2.2.   SOLAR ENERGY

Solar energy has emerged as one of the most promising technologies for alleviating Brazil's energy problems, where the heavy reliance on hydroelectric power stations has caused major problems. This has become evident during the period of low rainfall in recent years, which has caused an increase in electricity and water bills in much of the country, as well as rationing with few precedents (BLOG DA ENGENHARIA, 2016).

However, this scenario is set to change substantially over the next few years, thanks to technological and bureaucratic developments surrounding photovoltaic panels. ANEEL (the National Electricity Agency) has announced the approval of a measure that will

will bring major improvements to the government's plan to encourage the development of solar energy generation (BLOG DA ENGENHARIA, 2016).

Even though it has enormous potential, as shown in Figure 2, photovoltaic solar energy in Brazil was mainly used for small isolated or autonomous systems, which were installed in places where the electricity grid did not serve, generally isolated or difficult to access regions, which make it economically unviable to install distribution lines. Stand-alone systems are generally used for electrifying rural properties, pumping water, remote telecommunications centres and signalling systems.

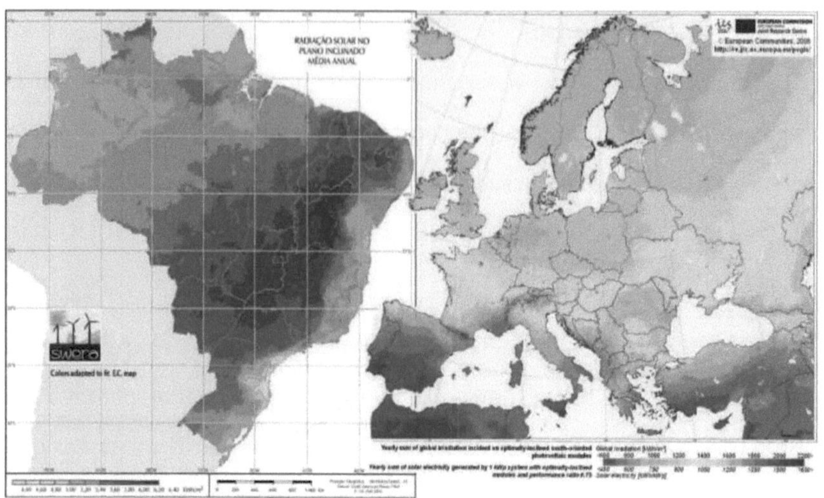

Figure 2 - Maps comparing solar irradiation values in Brazil and Europe.

Source: LabSol (2016)

With new regulatory resolutions, from now on, not only residential homes and buildings, but also businesses and even the industrial sector will be able to apply the so-called Compensation System, a model that basically allows the former customer of the energy concessionaire to **"sell" the energy generated through solar panels, receiving this value in** the form of a rebate on subsequent electricity bills (BLOG DA ENGENHARIA, 2016).

An important step towards the insertion of photovoltaic energy in the country was the **strategic** project **"Technical and Commercial Arrangements for the Insertion of Photovoltaic Solar Generation into the Brazilian Energy Matrix"**, launched by ANEEL IN 2011 in conjunction with companies and

electricity utilities throughout Brazil. The aim is to promote the creation of experimental photovoltaic power plants integrated into the national electricity system.

2.2.1. UTILISATION POTENTIAL

Compared to wind energy, photovoltaic solar energy is more regular in its

10

electricity supply, not to mention the fact that it can be used anywhere in Brazil, since our country is very privileged with high solar irradiation rates in all its regions.

The Northeast and Centre-West regions have the greatest potential for harnessing solar energy in the country. However, the other regions are not far apart in terms of insolation values. In Brazil, the South is the least favoured region in terms of insolation, but it still has better insolation than countries like Germany, which widely uses photovoltaic solar energy.

Germany is currently the country that employs the most photovoltaic energy in the world. With an installed capacity of around 20 GW, this is more than all the other countries combined. This is despite the fact that Germany has insolation of around 3500 Wh/m² (watt-hours per square metre) per day, available only in the south of its territory. Compared to Brazil, which has daily insolation values of between 4,500 and 6,000 Wh/m2.

Taking into account the size of our territory and our high irradiation rates, it is reasonable to think that our country has a photovoltaic generation potential at least ten times greater than Germany's current installed capacity. This would represent 200 GW of electricity from sunlight, more than all the electricity produced in the world in 2014, as can be seen in Graph 2.

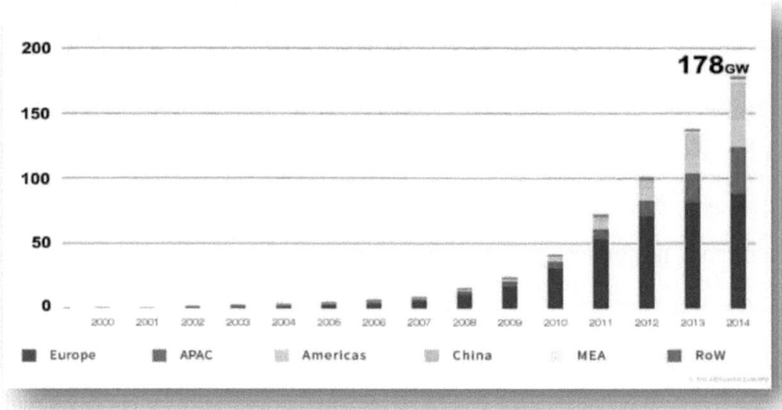

Graph 2 - Global installed capacity of photovoltaic energy in 2014.

Source: Solar Portal (2016).

There is therefore hope that with Brazil's immense potential for photovoltaic energy generation, it could become, if not the world's leading country, then one of the world's leading countries in the generation and use of alternative renewable energies (VILLALVA; GAZOLI, 2012).

2.2.2. OBSTACLES

As has happened with other energy matrices from alternative and renewable sources, such as wind energy, actions are expected to promote the insertion of photovoltaic energy in Brazil. Photovoltaics were left out of PROINFA, a programme created by the federal government to promote the use of alternative energy sources. It was also left out of the Ministry of Mines and Energy's Ten-Year Energy Plan until 2020.

Well valued in the most developed countries, photovoltaic energy was left aside for many years in Brazil, a country that has an abundance of sunlight throughout its territory, but little had been invested in photovoltaic energy until recently.

Several factors have contributed and some still contribute to the low use of this energy matrix in our country. Until 2012, the main obstacle was the lack of regulations and technical standards for the sector.

The cost of electricity generated with photovoltaic energy is still considered high when compared to hydroelectric energy and this is seen as a negative factor for the insertion of electricity in the country. However, this obstacle is practically non-existent for micro and mini photovoltaic plants installed in urban areas, where the cost of energy is higher due to taxes and transmission and distribution costs in the final price paid by the consumer. It can therefore be said that photovoltaic solar energy is economically viable and very competitive in the face of the high cost of electricity for Brazilian consumers (VILLALVA; GAZOLI, 2012).

1.1.1. BENEFITS

There are many advantages when we consider photovoltaic solar energy and all that it offers. There are not only economic benefits, but because it is a renewable source, it also benefits the environment. It's a constant source, meaning that even when you can't make use of the sun's energy, due to the night or cloudy and rainy days, you can count on it coming back the next day, which is why it's considered the most consistent and predictable of renewable energies.In addition, there are dozens of other benefits of photovoltaic solar energy, such as no noise when producing energy, the system is easy to install, there's little need for maintenance and the photovoltaic panels last more than 25 years.

1.1.2. REGULATIONS

2011 was a remarkable year in terms of progress in the photovoltaic solar energy sector in Brazil, mainly due to the results of the discussions generated by the Photovoltaic Energy Sector Group of the Brazilian Electrical and Electronics Industry

Association (ABINEE) and the CE-03:082.01 study commission of the Brazilian Electricity, Electronics, Lighting and Telecommunications Committee (COBEI), responsible for drawing up the standard for connecting photovoltaic inverters to the electricity grid.

In April 2012, ANEEL approved draft normative resolution 482, which allows micro-generation and mini-generation of electricity from renewable and alternative sources with a distributed generation system connected to low-voltage electricity grids. The resolution also establishes that each citizen or businessman can have a photovoltaic plant producing electricity to complement their own consumption or even to export energy, in the form of a system for offsetting electricity credits for self-producers of energy. (VILLALVA; GAZOLI, 2012)

2.3. BASIC CONCEPTS

This chapter will cover some important concepts that will help you understand how photovoltaic solar energy works, as well as its characteristics and properties.

2.3.1. RADIATION

It is the way in which solar energy is transmitted to our planet through space. Radiation is made up of electromagnetic waves that have different frequencies and wavelengths. Light travels at a constant speed in a vacuum. Equation I shows the relationship between frequency (f) in Hz, wavelength (X) in metres and the speed of light in a vacuum (c), where $c = 3 \times 10^8$ m/s.

$$c = \lambda . f \tag{I}$$

Radiation plays a direct role in the photovoltaic effect, which is the basis of photovoltaic solar energy systems for electricity production, which consists of transforming the sun's electromagnetic radiation into electrical energy by creating a potential difference, or an electrical voltage, on a cell formed by a sandwich of semiconductor materials.

Maranhão is a privileged state when it comes to generating solar energy. This is due to its geographical location, which means that the state has high levels of radiation throughout the year. Table I shows the results of the average daily solar irradiation in kWh/m² .day in three cities in the state of Maranhão.

Cities	Average daily solar irradiation (kWh /m² . day) 2015												
	Jan	Feb	Sea	Apr	May	June	Jul	Aug	Set	Out	Nov	Ten	Docto r.
São Luís	4,33	4,28	4,06	3,89	4,44	4,75	5,31	5,89	5,78	6,03	5,03	5,17	4,91
Turiaçu	4,06	4,56	3,94	4,03	4,39	4,61	4,97	5,61	5,58	5,72	5,72	5,00	4,85
Chapadi- nha	4,42	4,81	4,67	4,56	5,25	5,03	5,03	6,22	6,06	5,92	5,92	5,33	5,33

Table 1 - Average daily solar irradiation in three cities in Maranhão.

Source: Reference Centre for Solar and Wind Energy (2016).

Using the data from Table 1, Graph 3 is plotted, in which we can better observe the behaviour of the irradiation rates in the state of Maranhão in 2015, as well as the interference of the rainy season on the irradiation rates of the three cities.

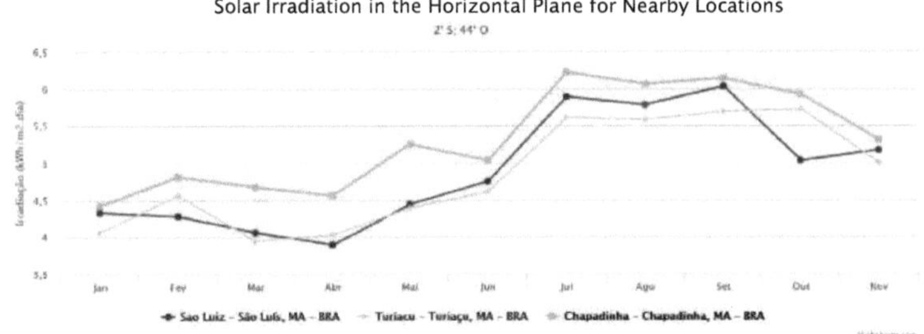

Graph 3 - Solar irradiation in three cities in Maranhão.

Source: Plotted using CRESESB tool

The quantity used to quantify solar radiation is irradiance, also known as irradiation, expressed as W/m² . This quantity can be measured using sensors and meters such as the one shown in Figure 3. On the Earth's surface, the irradiance of sunlight is typically around 1000 W/m2. This value is adopted as the standard in the photovoltaic industry for specifications and evaluation of photovoltaic cells and modules.

Figure 3 - PCE-SPM 1 solar irradiance meter

Source: PCE - IBERICA (2016).

2.3.2. INSOLATION

This is a quantity used to express the solar energy that falls on a given flat surface area over a given period of time. Its unit is Wh/m^2 (watt-hours per square metre). It expresses the density of energy per area (VILLALVA; GAZOLI, 2012).

Measuring insolation will be very useful when sizing photovoltaic systems, and in practice we find insolation tables and maps that provide daily values expressed in Wh/m2.

2.3.3. PHOTOVOLTAIC EFFECT

The capture of solar heat is the transformation of electromagnetic energy into thermal energy by the bodies and materials that receive its radiation. And when electromagnetic waves hit a certain body that has the capacity to absorb radiation, the electromagnetic energy can be transformed into kinetic energy and transmitted to the molecules and atoms that make up that body (VILLALVA; GAZOLI, 2012).The photovoltaic effect is called solar energy that is converted into electrical energy, through the incidence of solar radiation on one or more photovoltaic cells, the photons

16

that are part of it collide with the electrons present in the silicon structure, providing them with energy. And due to the electric field formed inside each cell, the electrons are forced to flow from the **P** layer to the **N** layer, thus generating a flow of electrons, an illustration of which can be seen in Figure 4.

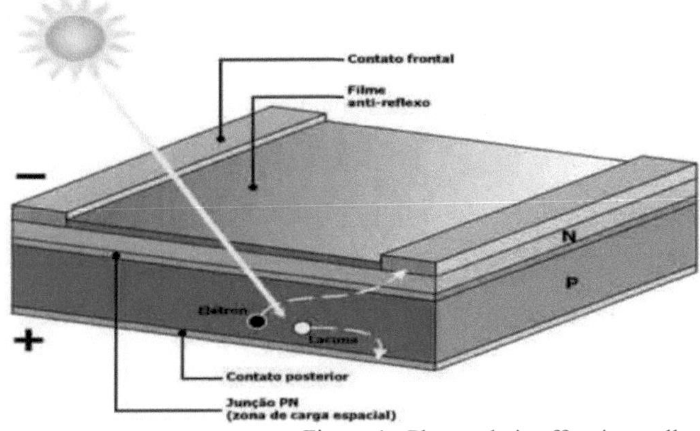

Figure 4 - Photovoltaic effect in a cell

Source: Apen Renewable Energies

Because each photovoltaic module is made up of a certain number of cells connected in series, when we join the negative layer of one cell with the positive layer of the next, electrical energy is produced.

2.3.4. PHOTOVOLTAIC MODULE ORIENTATIONS

The way in which the sun's rays reach the Earth directly affects the way in which modules should be installed, so it is necessary to have knowledge of the way in which these rays fall on our planet, using this knowledge to obtain better efficiency in capturing them.

It is also necessary to understand the azimuthal angle, which is the angle at which the sun's rays are orientated in relation to geographic north; as the sun has its own trajectory in the sky, the azimuthal angle is different throughout the day.

The correct installation of a photovoltaic solar module must take into account the movement of the sun, otherwise we won't get the expected results at any given time

17

of day. The best way to install a fixed solar module, without a solar tracking system, in the southern hemisphere is to orient it with its face to the geographical north, and in the northern hemisphere with its face to the geographical south, so that the sunlight is maximised.

2.3.5. CHOOSING THE ANGLE OF INCLINATION

Most photovoltaic systems have a fixed angle of inclination. This angle must be chosen on the basis of a number of criteria: geographical location, incidence of sunlight, etc. The incorrect choice will lead to a reduction in the amount of sunlight captured and compromise the module's electricity production. Naturally, with a fixed tilt module you can't maximise the amount of sunlight captured every day or month, but it is possible to choose an angle that provides a good average energy production throughout the year, the consequence of the different tilt angles is shown in Figure 5.

Figure 5 - Different angles of inclination of solar panels.

Source: BLOGVOLTAICO (2016).

There is no standard angle of inclination for installing solar modules, but it is possible to determine for a geographical latitude an angle of inclination that enables good average energy production throughout the year, as shown in Table 2.

Geographical latitude of the location	Recommended angle of inclination
0° to 10°	$\alpha = 10°$
11° a 20°	$\alpha = $ latitude
21° a 30°	$\alpha = $ latitude + 5°
31° a 40°	$\alpha = $ latitude +10°
41° or more	$\alpha = $ latitude +15°

Table 2 - Angle of inclination in relation to latitude.

2.3.6. BASIC RULES FOR INSTALLING SOLAR MODULES

There are basically two basic rules for installing solar modules correctly:

I - Whenever possible, orient the module with its face towards geographical north, as this maximises average energy production.

II - Set the correct angle of inclination of the module with respect to the ground in order to optimise energy production throughout the year. To do this, choose the angle of inclination according to the table.

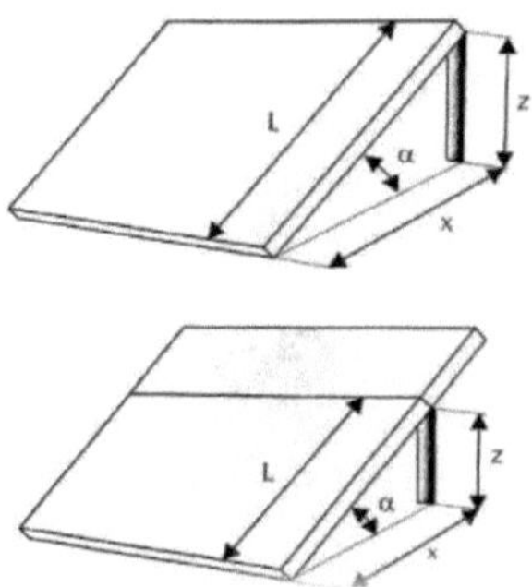

Figure 6 - Height and support rod of the module determine the angle of inclination

Source: VILLALVA, GAZOLI (2012).

Following rule II, **the value of the angle of inclination a** shown in Figure 6 is obtained. In practice, for physical installation, the height of the clamping rod (z) must be calculated as a function of **the calculated angle (α) and taking into account the length of the module (L).** The height z of the clamping rod is calculated using Equation 2:

$$z = L \cdot sen\ \alpha \tag{2}$$

And the distance x is calculated using Equation 3:

$$x = L \cdot cos\ \alpha \tag{3}$$

2.3.7. AUTOMATIC SUN POSITION TRACKING

Photovoltaic systems with automatic sun position tracking optimise the angle of incidence of the sun's rays automatically throughout the day, month and year. The system can have one or two degrees of freedom.

The aim of this work is to develop an experimental methodology for carrying out tests and developing hardware. The construction of this prototype of a photovoltaic solar module with an automatically optimised position will use Arduinos, a protoboard, light sensors and servo motors.

The ARDUINO platform website states that **"An Arduino is a** single board **microcontroller** with a software suite to programme it. The hardware consists of a simple free hardware design for the controller, with an Atmel AVR processor and inbuilt input/output support. The software consists of a standard programming language and the **bootloader that runs on the board."**

In practical terms, an Arduino is a small computer that can be programmed to process inputs and outputs between the device and the external components connected to it. It has a multitude of applications, from the simplest such as turning on a light and then turning it off after a certain period of time, to more advanced applications using a multitude of sensors, embedded, connected to a network, or even to the Internet to send a set of data received from some sensors to a website, which can then be displayed in the form of a graph (McRoberts, Arduino Basics).To do this programming, you use the Arduino IDE, a free software program in which you write the codes in the language recognised by the device (based on the C language). Both the hardware and the software are open source, meaning that the codes, schematics, design, etc. can be used freely by anyone and with any device. Because of this, there are a multitude of clone boards available for sale, or even ones that can be created from a diagram, but as the projects are open source, any board is 100% compatible with Arduino, just as any software, hardware, shield, etc, will also be 100% compatible.

2.4. PHOTOVOLTAIC ENERGY

The photovoltaic effect is the physical phenomenon that allows the direct conversion of light into electricity. This phenomenon occurs when light, or electromagnetic radiation from the sun, falls on a cell made up of semiconductor materials with specific properties. The semiconductor layers of the cell can be made from several different materials, the most common being silicon. Around 95 per cent of all photovoltaic cells manufactured in the world are made of silicon, which is an abundant and cheap material. (PHOTOVOLTAIC SOLAR ENERGY, MARCELO GRADELLA and JONAS RAFAEL)

2.4.1. PHOTOVOLTAIC MODULES AND PANELS

The output voltage and current of a photovoltaic cell are low, which is why several are grouped together to form a set of cells connected in series or parallel, depending on the voltage and current requirements.

According to Villalva and Gazoli (2012), a commercially available packaged set of cells is called a module, board or panel. In this work, the term used will be module. The set of modules connected in series or parallel to increase voltage or current is called a string.

The most common photovoltaic modules available on the market are made of crystalline silicon and produce between 50 and 250 W of power, with maximum voltages of approximately 37 V and can supply around 8 A of electric current (CRESESB, 2008).

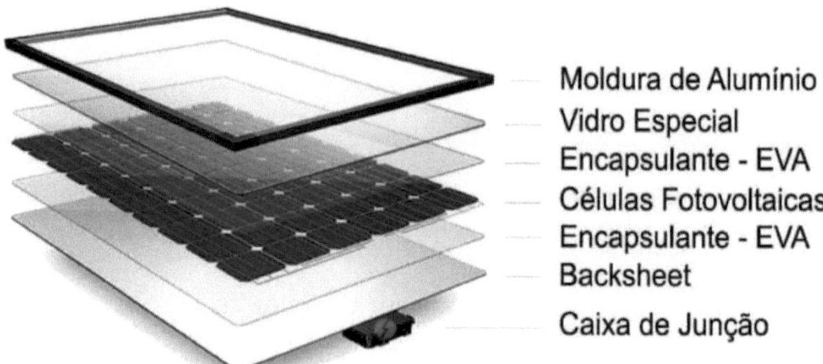

Figure 7 - Components of a photovoltaic module

Source: Solar Portal (2016).

Photovoltaic panels, such as the one shown in Figure 7, are made up of a group of electrically connected cells. One photovoltaic cell provides an electrical voltage of approximately 0.6 V. To produce modules with higher output voltages, manufacturers connect several cells in series. Modules usually have 36, 54 or 60 cells, depending on their power class.The electric current produced by a cell depends on its area, because the electric current depends directly on the amount of light received by the cell. The greater the area, the greater the light capture and consequently the greater the current supplied. Commercial crystalline modules generally supply around 8 A of electric current.

2.4.2. CHARACTERISTICS OF COMMERCIAL PHOTOVOLTAIC MODULES

In general, the power of modules is given by peak power, expressed in watt peak (Wp). However, there are other characteristics that better characterise the module's functionality. The main ones are (CRESESB, 2004):

- Open Circuit Voltage (Voc): maximum voltage that a device can deliver under certain radiation and temperature conditions, corresponding to zero current flow and, consequently, zero power.

- Short Circuit Current (Isc): the maximum current that a device can deliver under certain radiation and temperature conditions, corresponding to zero voltage and, consequently, zero power;
- Maximum Power (Pm): This is the maximum power value that the device can deliver. It corresponds to the point on the curve where the current versus voltage product (IxV) is maximum;
- Maximum Power Voltage (Vmp): This is the only voltage at which the maximum power can be extracted;
- Maximum Power Current(Imp): Current that the device delivers for maximum power under given radiation and temperature conditions.

2.4.3. CHARACTERISTICS OF THE CURRENT VERSUS VOLTAGE (IxV) CURVE.

When a photovoltaic cell is connected, voltage and current measurements can be plotted on graphs. To do this, all you need to do is vary the load conditions, as new voltage and current values will emerge. These values, when plotted on the same graph and connected by a line, give rise to the so-called IxV Characteristic Curve, as shown in Figure 8.

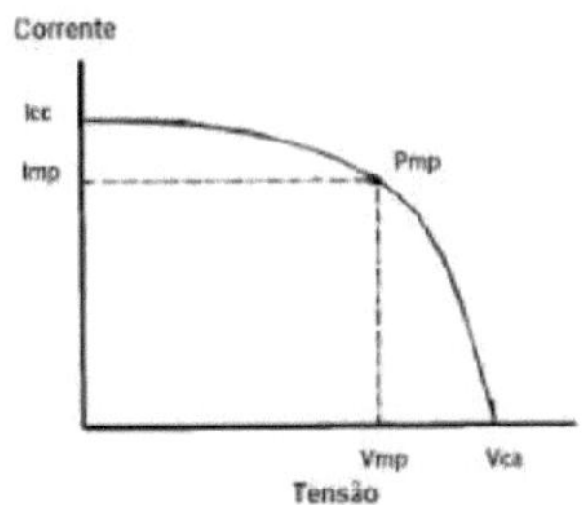

Figure 8 - Typical IxV characteristic curve of a photovoltaic module

Source: Engineering Manual for Photovoltaic Systems (2016).

For each point on the curve, the current-voltage product represents the power generated to determine the operating conditions. Figure 9 shows that for a photovoltaic cell there is only one voltage and corresponding current for which the maximum power can be extracted.

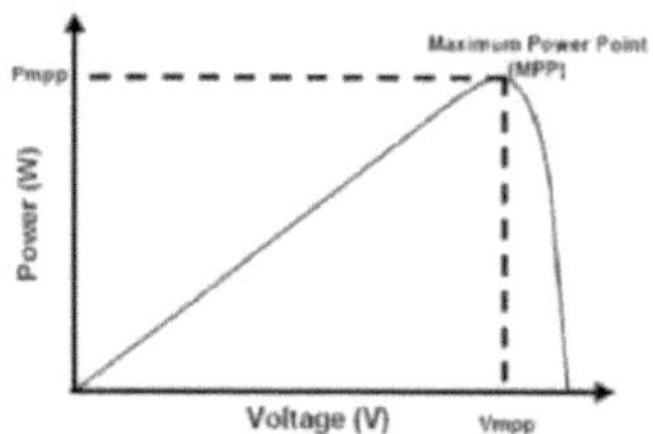

Figure 9 - Typical PxV curve for a photovoltaic module

Source: Engineering Manual for Photovoltaic Systems (2016).

2.4.4. FACTORS AFFECTING THE ELECTRICAL CHARACTERISTICS OF PHOTOVOLTAIC MODULES

There are a number of factors that influence the characteristics of a panel:

2.4.4.1. LIGHT INTENSITY

It is necessary to monitor the movement of the sun throughout the day in order to obtain greater light intensity. But one barrier is that the modules are generally installed in a fixed position, due to the high cost of the system's automation equipment. It is therefore essential to determine the best inclination for each region depending on the latitude and characteristics of the demand.

2.4.4.2. CELL TEMPERATURE

With an increase in the level of insolation, there is a consequent rise in the cell's temperature, causing a decrease in the module's efficiency. There are modules, such as amorphous silicon, which suffer less influence from temperature on peak power, although they also suffer a reduction in performance.

2.5. APPLICATION OF A PHOTOVOLTAIC SYSTEM CONNECTED TO THE ELECTRICITY GRID

A grid-connected photovoltaic system operates in parallel with the local electricity grid. The aim of this type of system is to generate electricity for local consumption, and it can reduce or eliminate consumption from the public grid or even generate surplus energy, thus ceasing to be consumers and becoming producers. The energy is sold on the commercial market and tariffs and other standardised technical requirements apply for this type of connection. In the electricity compensation system there is so-called **net metering**, where there is an electronic meter that registers the energy that the household consumes from the public grid and the energy that the household produces and eventually exports to the grid.

2.5.1. MAIN ELEMENTS OF THE SYSTEM

2.5.1.1.PHOTOVOLTAIC PANEL

Photovoltaic cells can produce small voltages in the 0.5 V range. Since most photovoltaic systems require a nominal voltage of 12 Volts, it is necessary to couple the cells in series. The structure in which this coupling is carried out is called a photovoltaic panel. Figure 10 shows the coupling of cells that form modules, which together form solar panels.

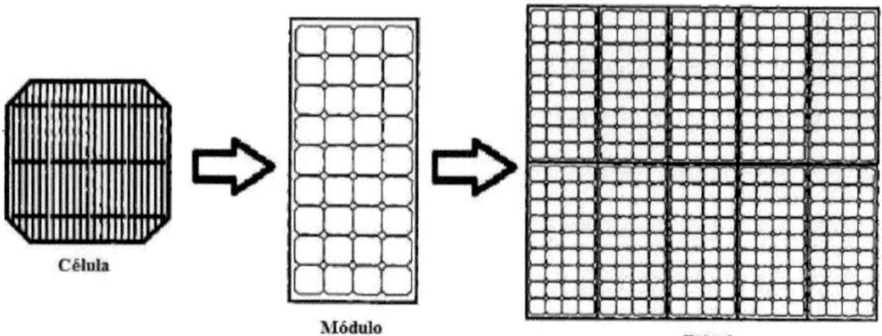

Figure 10 - Photovoltaic cell, module and panel.

Source: Control and Automation Engineering UFOP (2015)

The number of cells in a panel, and therefore its nominal output voltage, depends on the crystal structure of the semiconductor used. The manufacturer must decide on the number of cells taking into account the most unfavourable climatic situation. On the back of the panel are the electrical contacts. These must be protected to guarantee connectivity (GASQUET, 1997). Photovoltaic panels generally have a very similar ratio between the output current and the voltage supplied. This relationship, as seen above, is called the panel's IxV curve, meaning that photovoltaic panels have very similar IxV curves (GASQUET, 1997).

2.5.1.2.   INVERTER

As we have seen so far, photovoltaic panels are only capable of supplying current in the form of direct current (DC). In some applications it is possible to harness this current, but in many cases it is necessary to convert this current into an alternating current (AC) source.

The device responsible for transforming direct current into alternating current is called an inverter. This device is capable of transforming a 12 Vdc voltage source into a 110 VAC, 220 VAC source with frequencies of 50 or 60 Hz, or other combinations that may be interesting for the system. This conversion, however, involves losses

In other words, inverters have efficiencies of between 75 and 91 per cent. This is due to the fact that the consumption of the inverter circuit increases proportionally with the increase in the power it is controlling (GASQUET, 1997).

According to Pereira and Oliveira (2011) the inverters used to connect to the electricity grid have grid voltage control, output waveform control, synchronisation between the signal generated and the electricity grid and protection devices. These inverters, which convert direct current into alternating current, are usually called **grid-tie inverters** and are different from the **off-grid inverters** used in stand-alone systems, because the inverters for stand-alone systems supply electrical voltage and the inverters for connection to the grid supply electrical current and do not have the capacity to supply voltage to consumers; the voltage is the same as the grid to which the system is connected.

For safety reasons, grid-connected inverters only work when they are connected to a power grid, because if there is a disconnection or failure in the grid supply, the system should not supply power to the grid, providing safety for equipment connected to the same grid and for people who may be doing maintenance. The inverters should preferably be pure sine wave, i.e. they are pure wave inverters, which obtain top quality conventional electricity. There are also other system components such as the string box, which connects the strings in the set. The DC protection board and AC protection board, which are used to disconnect the generation system from the consumer. The distribution board receives the energy from the grid or the photovoltaic system and distributes it for consumption. The meter is the device responsible for measuring the amount of energy that has been injected into the grid and the energy consumed from the grid. It differs from traditional meters, which only record the energy consumed.

2.5.2. PROTOTYPE ELEMENTS

The development of a robotic device (prototype) will help to understand how an autonomous tracking system works, making it possible to make better use of the solar energy captured by photovoltaic panels. It is based on an experimental methodology that will guide the tests needed to develop it. Initially, the prototype will require an Arduino board, protoboard, light sensors and servo motors, such as those shown in Figure 11.

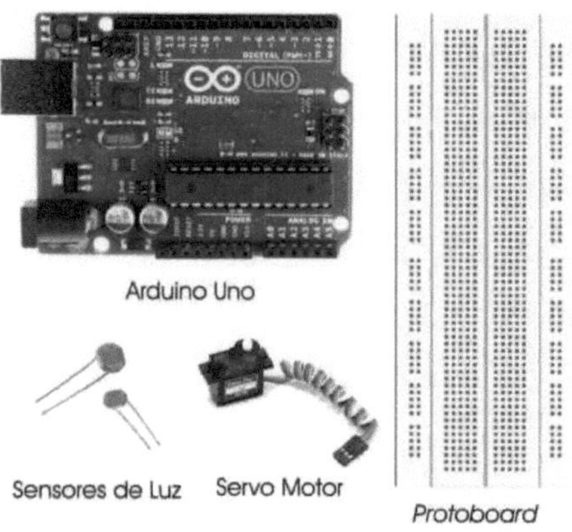

Figure 11- Main components for making the prototype

Source: Arduino extended summary (2015).

The purpose of this prototype is to ensure that the solar plate or panel is always aligned perpendicular to the sun's rays in an autonomous way, thus providing a system with a power gain. To do this, it is necessary to use the Light Dependent Resistor (LDR) sensors to make the system realise the actual position of the incident rays.

Two servo motors will be used in this prototype, one responsible for the tilt movement and the other for the azimuth angle, both independent of each other and/or able to act simultaneously.

CHAPTER 3

RESULTS

In this section, new concepts will be used, as well as those already covered, to simply dimension a photovoltaic solar energy generation system that will meet the electricity needs of two low monthly consumption households. The aim is to advance the understanding and sizing of this type of system, so that in the future one can be realised for the new BICT building at the Federal University of Maranhão.

3.1. DIMENSIONING A SYSTEM

The aim of this work is to carry out a theoretical study for the future preliminary sizing of an auxiliary electricity generation system made up of solar panels for the new BICT building. To do this, it will be necessary to carry out a study on the applications of a mini-generation system shared between two homes that have a lower demand for energy generation compared to the new BICT building.

The modules can be connected in series or in parallel to create a set with greater energy supply capacity and higher voltages and currents than those produced by an individual panel. When connected in series, the modules form rows or strings, and the number of modules connected in series determines the voltage of the photovoltaic array, which is applied to the input terminals of the inverter, which needs to be sized to withstand the sum of the open circuit voltages of the modules.One fact that must be taken into consideration is the choice of electrical cables, since they provide access to the electricity produced in the modules. They are usually supplied by the manufacturers because they have specific characteristics for applications in this segment. Photovoltaic systems generally work with higher direct current voltages than the alternating current voltages of conventional electrical installations, in addition to exposure to excessive solar radiation, which ends up drying out cables not designed for this type of application.

3.2. ENERGY PRODUCED

The first decision to be made for a grid-connected system is to determine how much energy you want or need to produce, taking into account various criteria. In this case study, the energy to be produced by the photovoltaic system was determined based on the average monthly electricity consumption of a home located in the Parque Timbiras neighbourhood in the city of São Luís/Ma. This data was obtained from the home's electricity bills and the result is shown in Graph 4.

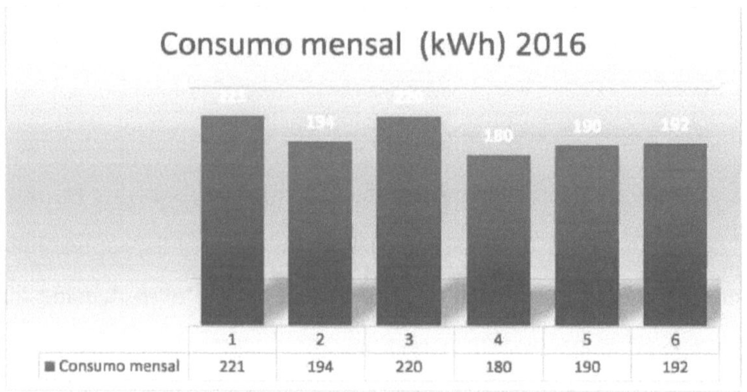

Graph 4 - History of electricity consumption over the last 6 months.

Source: The author (20160.

The aim of this system will be to fully meet the electricity demand of two homes with the same consumption as my home. Based on the history of monthly consumption, the average consumption was approximately 200 KWh. The system should therefore produce 400 kWh per month. And as an automatic tracking system will be used, there is no need to choose the appropriate azimuth and inclination angles, as the system itself will make the corrections when necessary.

3.3. DIMENSIONING THE NUMBER OF MODULES

Once you have chosen a module model to use, you need to determine how much you need for the system. All grid-connected systems have MPPT (maximum point tracking) systems, which are present in the inverters and serve to maximise the power supplied by the photovoltaic modules, making them always operate at their maximum power point, regardless of the conditions that affect performance and alter the current and voltage characteristic curve of the module set.

The most appropriate method was chosen based on daily insolation, i.e. the daily kilowatt-hour square metre value (kWh/m²) available in the location where the photovoltaic power generation system will be installed.

Once you know the area of the module and its efficiency, you can easily calculate the electricity produced daily by the module. And for monthly production, you just need to multiply the daily figure by thirty. But manufacturers already provide this data in technical sheets, tables, etc.

Under these conditions, the Bosch 240 W monocrystalline photovoltaic panel produces 37 kWh/month, knowing the amount of energy you want to produce each month. The next step in sizing the system is to determine the number of modules needed in the photovoltaic system, using Equation 4:

$$N_P = \frac{E_{SISTEMA}}{E_{MODULE}} \tag{4}$$

Where:

N_P = number of modules in the photovoltaic system

E_{SYSTEM} = energy produced by the system (kWh) in a certain time interval

E_{MODULE} = energy produced by the module (kWh) in a certain time interval

$$N_P = \frac{400\ kWh}{37 kWh}$$

$N_P \approx 11$ modules

3.4. INVERTER SIZING

In order to choose the type of inverter to be used in the system, certain criteria must be taken into account:

- The string open-circuit voltage must not exceed the maximum permissible voltage at the inverter input. Otherwise the inverter may be irreversibly damaged.
- The inverter must be specified for a power equal to or greater than the peak power of the module set.

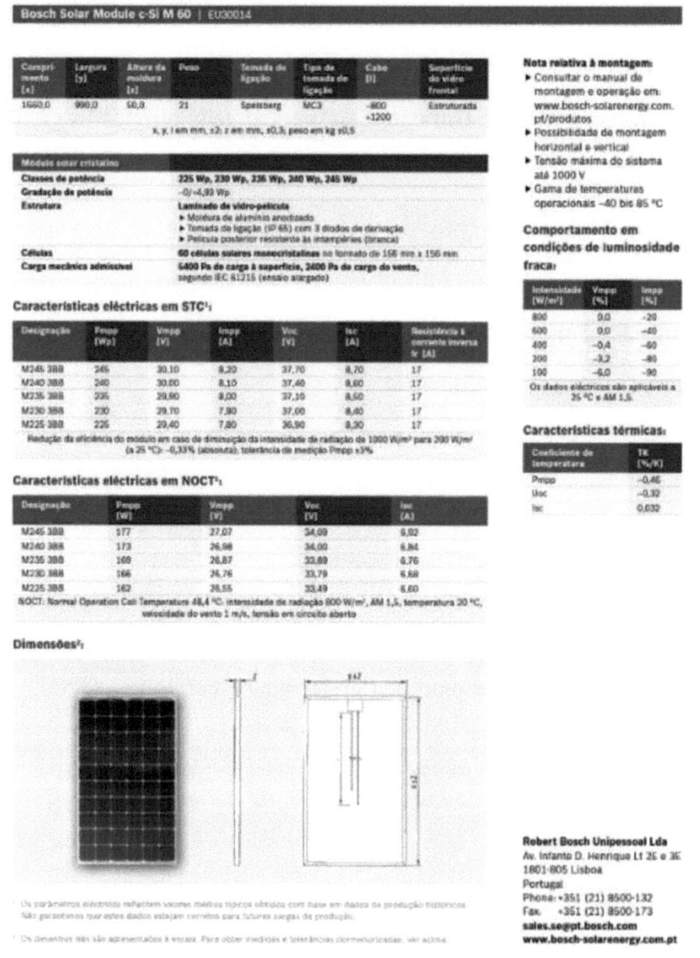

Figure 12 - data sheet for Bosch monocrystalline modules

Checking the Bosch data sheet shown in Figure 12, we see that the open circuit voltage of the modules under Standard Test Conditions (STC) is **V_{oc} = 37.4 V.**

With 11 modules connected in series, the string's maximum output voltage is given by Equation 5:

$$V_{OC/STRING} = 11 \times 37.4 \text{ V} \qquad (5)$$

$$= 411{,}4 \text{ V}$$

Considering 10% as a standard safety factor, the maximum output voltage of the string is given by Equation 6:

$$V_{OC/STRING} = \textbf{411.4x0.1} \qquad (6)$$

$$V_{OC/STRING} = \textbf{452.54 V}$$

To be sure of the open-circuit voltage that will be found at the output of the string, the temperature coefficient specified in the module's data sheet was used. The temperature coefficient for the open circuit voltage was found to be -0.32%/k, i.e. for each degree of temperature reduction there will be a 0.32% increase in the module's output voltage. Considering that the operating temperature in São Luís/Ma will always be above 5°C, we calculated:

$$\text{Percentage voltage variation} = (26 - 5) \times 0.32\% = 6.7\% \qquad (7)$$

$$\text{Voltage variation} = 6.7\% \times 411.4 \text{ V} \qquad (8)$$

$$= 27{,}6 \text{ V}$$

So there you have it:

$$\textbf{V}_{OC/sTRiNG} = 411.4 \text{ V} + 27.6 \text{ V}$$

$$= 439{,}0 \text{ V}$$

It can be seen that the rule of thumb of using a safety factor of 10 per cent results in a higher, and therefore safer, value than the calculation using the temperature coefficient.

It is then necessary to find a suitable inverter line that supports the connection of the modules in series with a total voltage of 452.54 V. An example of an inverter suitable for use in this system is the Santerno M PLUS inverter. The data sheet for this inverter states that the maximum continuous voltage supported is 600 V. Now all you have to do is choose an inverter model from this range that is compatible with the power of the modules, given by Equation 9:

$$P_{MODULE} = 11 \times 240 \text{ W} \tag{9}$$

$$= 2640 \text{ W} = 2.64 \text{ kW}$$

Searching for information in the Santerno manufacturer's catalogue, you will find the M PLUS 3600 model, which supports up to 3.3 kW at its direct current input, making it suitable for application in this project.

Then the system:

It will produce 400 KWh per month;

It will have 11 Bosch 240 W monocrystalline photovoltaic modules;

With a 10% safety factor, the string's maximum output voltage will be 452.54 V;

And module power equal to 2640 W or 2.64 KW;

Santerno's M PLUS 3600 inverter.

A quick price search was carried out on websites that sell photovoltaic system equipment, and the results are shown in Table 3.

EQUIPMENT	Quantity	Market value.	Value of investment in our system
Photovoltaic module	11	R$ 1.300,00	R$ 14.300,00
Inverter	1	R$ 8.000,00	R$ 8.000,00

Table 3 - Equipment values.

Source: The author (2016).

The Solar Portal simulator was used to calculate the average price of a photovoltaic solar energy system with identical characteristics to the one dimensioned in this work. The simulator showed that a photovoltaic solar energy system previously dimensioned requires an investment of between R$27,000.00 and R$31,000.00, with all the components already installed.

A study was carried out on the cost of the kWh paid in São Luís do Maranhão, based on the amount charged in June 2016. Under these conditions, it would take us 10.8 years to pay for a photovoltaic energy generation system, just with the money we would spend on energy bills. Taking into account that a photovoltaic energy system has a useful life of more than 25 years, the energy generation system becomes a good investment.

CHAPTER 4

FINAL CONSIDERATIONS

Based on the study carried out, it can be seen that photovoltaic solar energy is very efficient when it comes to clean, renewable energy. Even though the cost of implementation is still considered high, this type of system is viable due to the low need for maintenance and the return on investment in a short period of time.

Given the importance of developing new energy sources, especially when it comes to renewable and clean energies, and in view of the great energy potential that exists in São Luís do Maranhão. This city receives a large amount of solar radiation throughout the year, and the application of a photovoltaic solar energy generation system will bring great benefits not only financially, but also in the academic context for the Federal University of Maranhão. Noting that automation systems are always undergoing constant changes and updates, it could serve as a field of research for students and researchers.

As this is a technology that has only recently been used on a large scale, photovoltaic solar panels are still the subject of research to improve their performance. In order to maximise the efficiency of this equipment, it is necessary to develop an automation system based on the development of a prototype system for automatically tracking the position of the sun. This prototype is intended to be presented in the Contextualisation and Curricular Integration Work II (TCIC II), along with the sizing of a system suitable for the BICT building at the Federal University of Maranhão (UFMA), taking into account the area, insulation and various other factors in the region of the new building.

This work is therefore a preliminary study, which will continue with the acquisition of equipment, the development of a prototype, the search for the architectural design and plans for the building, and other information relevant to the project.

CHAPTER 5

BIBLIOGRAPHY

ALVES, A. F. **Development of an Altomatic Positioning System for Photovoltaic Panels.** Faculty of Agronomic Sciences, UNESP, Botucatu, SP, 2008.

ANEEL. National Electricity Agency. **Brazil's Electricity Highs: Part 2 Renewable Sources.** Brasilia 2007.

ANEEL. National Electricity Agency. NORMATIVE RESOLUTION NO. 482, OF 17 APRIL 2012. Available at < http://www2.aneel.gov.br/cedoc/ren2012482.pdf>

ARDUINO. **About the Arduino Platform.** Available at <http://www.arduino.cc/>.

ArduinoLabs (2012). Available at: <http://arduinolabs.in/girasol-siga-a-luz/>

ENGINEERING BLOG. **SOLA ENERGY.** Available at: <http://blogdaengenharia.com/geracaodomesticadeenergiasolarficamenosburocraticae pode- servendida/ > accessed on 22/06/2016

BORGES NETO, Manuel Rangel de; CARVALHO, Paulo. **Electricity Generation: Fundamentals.** São Paulo: Érica, 2014.

CEPEL/CRESESB - Electric Energy Research Centre. Sérgio de Salvo Brito Reference Centre for Solar and Wind Energy. **Engineering Manual for Photovoltaic Systems.** Rio de Janeiro: 1999.

CEPEL/CRESESB. **Engineering Manual for Photovoltaic Systems.** 2004. Available from: <http://www.cresesb.cepel.br/tutorial/tutorial_solar.htm>

CRESESB. **SOLAR ENERGY PRINCIPLES AND APPLICATIONS.** 2006.Available at: <http: http://www.cresesb.cepel.br/download/tutorial/tutorial_solar_2006.pdf >.

DAZCA, Rafael Guershom. **STUDY OF THE IMPLEMENTATION OF A**

PHOTOVOLTAIC SOLAR ENERGY SYSTEM IN A BUILDING AT THE MACKENZIE PRESBYTERIAN UNIVERSITY. Article - Mackenzie Presbyterian Institute, School of Engineering, Mechanical Engineering Coordination.

FOWLER, Michael. **THE PHOTOELECTRIC EFFECT.** available at < http://galileo.phys.virginia.edu/classes/252/photoelectric_effect.html>

GASQUET, Héctor L. PHOTOVOLTAIC SYSTEMS. 1997. El Paso, Texas. HOWSTUFFWORKS, available at < http://www.howstuffworks.com/solar-cell.htm>

HALMEMAN, RADAMES JULIANO. DEVELOPMENT OF A SYSTEM FOR REMOTE MONITORING OF PHOTOVOLTAIC MICROGENERATION PLANTS. Thesis (Doctoral Student) - Agronomy Course (Energy in Agriculture), Unesp - **Universidade Estadual Paulista "Júlio De Mesquita Filho", Botucatu, 2014.**

INCROPERA, Frank P. FUNDAMENTS OF HEAT AND MASS TRANSFER. LTC- Livros Técnicos e Científicos Editora S.A. Fourth edition, 1996. Rio de Janeiro, RJ.

JORNAL DO BRASIL. DEMANDA ENERGETICA NO BRASIL. Available at: <http://www.jb.com.br/economia/noticias/2014/02/23/demanda-por-energia-no-brasil-e- insustentavel/> Accessed on 19/06/2016.

LORENZO E. ELETRICIDADE SOLAR: ENGENHARIA DE SISTEMAS FOTOVOLTAICOS. Spain, Artes Gráficas Gala, S.L., 1994.

OGLOBO, BRAZIL FACES WORST ENERGY CRISIS IN HISTORY. Available at <http://noblat.oglobo.globo.com/geral/noticia/2015/01/brasil -enfrenta-pior- crise-energetica-da-historia.html> accessed on 19/06/2016

RESNICK, Halliday Walker. FUNDAMENTALS OF PHYSICS. LTC- Livros Técnicos e Científicos Editora S.A. Fourth edition, 2002. Rio de Janeiro, RJ.

VILLALVA, G. GAZOLI, J. ENERGIA SOLA FOTOVOLTAICA, CONCEITO E APLICACÕES. 1ed -São Paulo: Erica, 2012.

4.1. WEBGRAPHY

http://www.portalsolar.com.br/ (Portal with various information on photovoltaic energy)

http://www.cresesb.cepel.br (Reference Centre for Solar and Wind Energy)

<https://apenergiasrenovaveis.wordpress.com/solar/energia-solar-fotovoltaica-como-se- produces/> Accessed on 23/07/2016

yes I want morebooks!

Buy your books fast and straightforward online - at one of world's fastest growing online book stores! Environmentally sound due to Print-on-Demand technologies.

Buy your books online at
www.morebooks.shop

Kaufen Sie Ihre Bücher schnell und unkompliziert online – auf einer der am schnellsten wachsenden Buchhandelsplattformen weltweit! Dank Print-On-Demand umwelt- und ressourcenschonend produziert.

Bücher schneller online kaufen
www.morebooks.shop

Printed by Books on Demand GmbH, Norderstedt / Germany